DJI Air 3S

A Full Spectrum Flight Guide for Drone Enthusiasts and First-Time Flyers

Everything You Need to Know About Operating, Capturing, and Protecting Your Aerial Adventure

Joe E. Grayson

Copyright © 2024 Joe E. Grayson, All rights reserved.

No part of this publication may be reproduced, distributed, or transmitted in any form or by any means, including photocopying, recording, or other electronic or mechanical methods, without the prior written permission of the publisher, except in the case of brief quotations embodied in critical reviews and certain other noncommercial uses permitted by copyright law.

Table of Contents

Table of Contents
Introduction
Chapter 1: Unboxing and Setup
Chapter 2: Navigating the DJI Fly App
Chapter 3: Flight Modes and Controls
Chapter 4: Camera Settings and Filming Techniques
Chapter 5: Understanding GPS and Satellites
Chapter 6: Obstacle Avoidance and Safety Features
Chapter 7: Advanced Flight Techniques
Chapter 8: Drone Maintenance and Troubleshooting
Chapter 9: Flying Legally and Responsibly
Chapter 10: Troubleshooting and Fixing Errors
Chapter 11: The Future of Drone Technology
Conclusion

Introduction

The world of drones is evolving at an incredible pace, and at the forefront of this revolution stands DJI, a brand that has become synonymous with cutting-edge technology and exceptional quality. DJI has been a trailblazer in the drone industry, constantly setting the standard for innovation, reliability, and performance. From the earliest models to the sophisticated machines available today, DJI has continued to push boundaries, making drones accessible to enthusiasts and professionals alike.

Among its impressive lineup, the DJI Air 3S stands out as a true game-changer. Whether you're a seasoned pilot looking for a high-performance drone or a first-time flyer hoping to experience the thrill of flight, the DJI

Air 3S delivers a perfect blend of ease-of-use and advanced features. It's a drone designed to cater to a broad spectrum of users, from casual hobbyists to those who are passionate about capturing breathtaking aerial footage. What makes the Air 3S exceptional isn't just its technical specs or sleek design—it's the way it empowers users, offering a seamless flying experience that combines user-friendly controls with professional-grade results.

With its intuitive interface, powerful sensors, and advanced imaging capabilities, the DJI Air 3S transforms the way users approach drone flying. For beginners, it's a gateway to mastering flight techniques with minimal learning curve. For experienced pilots, it's a tool to elevate their creative projects, capturing stunning visuals that were once reserved for high-end, expensive drones. From the moment you take it out of the box, the DJI Air 3S invites you to experience a level of ease and performance that makes every flight a memorable adventure.

This guide has been created with you in mind—whether you're just starting your drone journey or you're looking to sharpen your skills. Its purpose is to give you the clearest, most comprehensive breakdown of the DJI Air 3S, ensuring that you feel confident in every aspect of operating the drone. We will explore its key features, provide practical tips for flying, and dive into the capabilities that set it apart from other models. Most importantly, we'll focus on making sure you understand how to safely navigate the skies and capture beautiful footage that will impress everyone who sees it.

Our goal is to make sure you not only understand the features and technical aspects of the DJI Air

3S but also know how to use them to their fullest potential. Whether you're looking to fly for fun, record your adventures, or create stunning video content, this guide will arm you with all the knowledge you need to take your drone flying to the next level.

Chapter 1: Unboxing and Setup

When you first unbox the DJI Air 3S, the excitement of finally owning this powerful drone is palpable. But before you take to the skies, it's important to familiarize yourself with the contents of the package to ensure you have everything you need to get started. The DJI Air 3S package is designed with ease of use in mind, providing all the essential components to help you hit the ground running. Here's a detailed look at what you'll find inside the box:

1. DJI Air 3S Drone Body

The centerpiece of the package is, of course, the DJI Air 3S drone itself. The drone body is lightweight yet robust, designed to withstand the rigors of flight while being easy to transport. Its sleek, compact form factor makes it a joy to handle, and the construction quality is immediately noticeable. The Air 3S features an advanced camera system with larger sensors that allow it to capture stunning visuals, whether you're flying for fun or working on a creative project. You'll also notice the redesigned propellers and the streamlined design that enhances the drone's aerodynamics and stability during flight.

2. RC2 Remote Controller

The RC2 remote controller is one of the standout features of the Air 3S package. It provides a smooth and intuitive flying experience with its responsive controls. The controller comes with a built-in screen that makes it easy to view live footage from the drone in real-time, without the need to attach your smartphone or tablet. It offers a more streamlined and integrated flying experience, eliminating the hassle of connecting external devices. The RC2 is equipped with high-quality joysticks, customizable buttons, and an ergonomic grip to ensure comfort during long flying sessions. Additionally, the controller features DJI's OcuSync 3.0 technology for long-range, stable connectivity with the drone.

3. Batteries

The DJI Air 3S comes with a high-capacity, rechargeable battery designed to provide extended flight times. Depending on the specific package you purchase, you might find one or two batteries included. These batteries are easy to insert and remove, and they come equipped with intelligent battery management systems to ensure safe usage. The Air 3S can deliver flight times of up to 46 minutes per charge, but this can vary based on weather conditions, altitude, and how the drone is being flown. Always ensure you have a spare battery or two on hand, especially for longer filming sessions or travel excursions.

4. Charging Hub and Cables

Included in the box is a charging hub that allows you to charge the drone's batteries efficiently. This hub can typically charge up to three batteries at a time, ensuring that you're always ready for your next flight. Along with the charging hub, you'll find the necessary charging cables to connect the batteries, remote controller, and other devices to a power source. DJI has made sure that charging your gear is as seamless as flying your drone, so the process is simple and straightforward.

5. Propellers

To ensure smooth and stable flight, the DJI Air 3S comes with a set of durable propellers. These propellers are designed to be easy to install and replace. The package typically includes a spare set, so you're prepared in case of any damage during flight. DJI has designed these propellers to be both lightweight and powerful, maximizing the drone's efficiency while ensuring quieter operation compared to older models.

6. ND Filters

The DJI Air 3S comes with a set of ND (Neutral Density) filters to help you capture perfect footage, even in bright conditions. These filters reduce the amount of light entering the camera sensor, allowing for better control over exposure settings. Whether you're filming during the day or in direct sunlight, these ND filters allow you to capture cinematic, smooth shots without overexposing the image. These filters can be easily attached to the camera lens and come in various strengths for different lighting conditions.

7. Gimbals & Camera Protection

Protecting the drone's camera system is key, especially during travel or storage. DJI includes a gimbal protector that helps safeguard the sensitive gimbal and camera components from damage. The gimbal itself, which stabilizes the camera to ensure smooth footage during flight, is precision-engineered and often comes with a protective cover.

8. User Manual & Quick Start Guide

Finally, included in the box is a user manual and a quick-start guide. These materials provide essential instructions for getting started with your new DJI Air 3S. While the drone is designed to be easy to use, especially for beginners, the manual will help you understand all of the features and functions in greater detail. It also offers troubleshooting tips and safety guidelines to ensure you're flying responsibly and within the legal boundaries.

9. Additional Accessories (Depending on the Package)

Depending on the specific package or bundle you choose, there may be other accessories included. These could range from carrying cases to custom filters or even a set of replacement parts like motor mounts or screws. Always check the contents of your specific package to make sure you've received all the items you'll need.

Once you've familiarized yourself with all the components in the DJI Air 3S package, you'll be ready to assemble and prepare the drone for its first flight. Each part has been designed with precision and ease of use in mind, ensuring that you can quickly and safely get started with your drone adventure. Now, let's dive deeper into each component, exploring how they work together to make the DJI Air 3S an exceptional flying machine.

Now that you've unboxed your DJI Air 3S, it's time to take the first steps toward getting your drone ready for flight. The initial setup process is straightforward, but it's crucial to follow each step carefully to ensure a smooth and safe experience. Here's a breakdown of the initial setup process:

1. Powering Up the Drone and Controller

Before anything else, make sure your drone and controller are charged and ready to go. The DJI Air 3S and its remote controller come with partially charged batteries, but it's always a good idea to fully charge both devices before starting.

To power up the drone:

- **Press and hold the power button** on the drone's battery for about 2-3 seconds. You should see the LED lights on the drone body begin to flash, indicating that the drone is powered on.
- **Power up the RC2 controller** by pressing and holding the power button on the controller for a couple of seconds. The screen should light up, and you'll be greeted with the DJI logo, signaling that the controller is ready.

At this point, both devices should be turned on and ready for the next step. If the controller is not connecting automatically to the drone, you may need to initiate a pairing process, which we'll cover shortly.

2. Installing the Propellers and Battery

One of the first tasks when setting up your DJI Air 3S is installing the propellers and the battery. Here's how to do it:

Installing the Propellers: The DJI Air 3S comes with a set of propellers that need to be securely installed before the first flight. The propellers are designed for easy installation, and each one has a specific location on the drone for optimal performance.

- Start by inspecting the propellers. The propellers should be installed with the curved side facing upwards (toward the drone).

- On the drone, you'll notice that each motor has a color-coded marking (either a red or black dot). Match the color of the propeller to the motor markings. The red markings on the propellers should be installed on the motors with the red dots, while the black markings go on the motors with the black dots.
- Gently attach the propellers by pressing down on them and twisting until they click into place. Make sure they are secure and that the propellers rotate freely without obstruction.

Installing the Battery: To install the battery:

- Slide the battery into the compartment at the rear of the drone. You'll feel a slight resistance as the battery connects to the power terminals inside the drone. Make sure the battery is fully inserted, and you'll hear a click when it locks into place.

- The battery will power up the drone, and you should see the indicator lights on the front of the drone illuminating, confirming that it is securely installed and ready for flight.

3. Connecting the Controller to the Drone via the DJI Fly App

The DJI Fly app is the key to controlling and monitoring your DJI Air 3S, offering an intuitive interface for flight controls, camera settings, and system status. The app also provides live video feeds, maps, and real-time telemetry data, allowing you to stay informed during your flight.

Here's how to connect the controller to the drone:

- **Install the DJI Fly app** on your mobile device (if you haven't done so already). The app is available for both iOS and Android devices, and you can download it from the respective app stores.
- **Power on the controller** and the drone, as previously mentioned.
- **Connect your mobile device** to the RC2 controller using the appropriate cable (usually a USB-C or Lightning cable depending on your device). The DJI Fly app should automatically launch on your phone once the connection is made.
- On the app's screen, you'll see a prompt to connect to the drone. The controller will automatically search for your drone, and

once it's found, it will establish a connection.
- If this is your first time connecting the controller to the drone, you may need to go through the **pairing process**:
    - Open the DJI Fly app and select the "Connect" button when prompted.
    - The controller will display a pairing code and the drone will emit a sound to indicate it is pairing with the remote.
    - Once the pairing process is complete, the drone's status should appear on the app, and you'll be able to control it through the on-screen controls.

Once you've connected the drone to the DJI Fly app, the app will likely notify you of any firmware updates that need to be installed. It's a good idea to keep your drone's firmware up to date to ensure that you're using the latest features and improvements. Here's how to check for updates:

- Open the DJI Fly app and navigate to the settings menu.
- If there's a firmware update available, the app will prompt you to download and install it.
- Make sure your drone and controller are connected to a stable Wi-Fi network before initiating the update. The process typically takes a few minutes, so ensure that both the drone's battery and the controller's battery are sufficiently charged.

Before you take off, it's important to perform a few calibration steps to ensure that the drone flies smoothly and safely. Some of these steps will be done automatically by the DJI Fly app when you first connect to the drone, but you should also perform a few manual checks:

- **Compass Calibration:** If the app prompts you to calibrate the compass, follow the on-screen instructions. This typically involves rotating the drone in specific directions to ensure that the compass is aligned correctly.
- **IMU Calibration:** The Inertial Measurement Unit (IMU) helps the drone understand its orientation in space. If needed, you'll be prompted to calibrate the IMU, which is important for stable flight.
- **GPS Signal:** Check that the drone has a strong GPS signal before taking off. The

app will display the number of satellites the drone is connected to—ideally, you want to have at least 10 satellites for a stable and safe flight.

Once the initial setup is complete, and the drone and controller are connected, you'll be ready for your first flight. However, always remember to conduct a few additional safety checks and familiarize yourself with the drone's settings, controls, and features in the DJI Fly app. This will help you make the most of your DJI Air 3S, whether you're capturing breathtaking aerial footage or simply enjoying a leisurely flight.

Chapter 2: Navigating the DJI Fly App

Getting started with the **DJI Fly app** is essential for a smooth and successful experience with your DJI Air 3S. The app serves as the central hub for controlling your drone, accessing its camera settings, viewing telemetry data, and much more. Here's how to get the DJI Fly app set up and a detailed look at its interface:

Before you can connect your DJI Air 3S to the controller and begin flying, you'll need to download and install the DJI Fly app on your mobile device.

Downloading the DJI Fly App:

1. **For iOS Devices:** Open the **App Store** on your iPhone or iPad and search for "DJI Fly." The official app will appear at the top

of the search results. Tap the "Get" button to begin downloading.

2. **For Android Devices:** Open the **Google Play Store** and search for "DJI Fly." Tap "Install" to download and install the app.

Once downloaded, you'll need to set up the app by following a few simple steps:

1. **Launch the App:** Open the DJI Fly app on your device.
2. **Sign in:** If you have an existing DJI account, you can sign in. If you're new to DJI, you'll need to create an account. This will allow you to access your drone's settings, software updates, and flight data.
3. **Agree to Terms and Conditions:** Make sure you read and agree to DJI's terms of service before continuing.

Once you have the DJI Fly app installed on your device, it's time to connect your **DJI Air 3S** and **controller** to the app.

1. **Connect the Mobile Device to the Remote Controller:** Using the appropriate cable (USB-C or Lightning, depending on your device), connect your phone to the controller's USB port. Make sure your controller is powered on.
2. **Power on the Drone:** Turn on the DJI Air 3S by holding down the power button on the battery for 2-3 seconds.
3. **Launch the DJI Fly App:** The app should automatically detect the drone and prompt you to connect. Tap on the "Go Fly" button in the app to begin the pairing process.
4. **Pairing:** If this is the first time connecting your controller to the drone, the app will

initiate the pairing process. Follow the on-screen instructions to complete the connection.

Once successfully paired, you will see the live video feed from your drone's camera on the app's main screen, and you'll be ready to begin exploring the features and controls.

The **DJI Fly app interface** is user-friendly and intuitive, making it easy for both beginners and experienced drone pilots to navigate and control the DJI Air 3S. Let's break down the key elements of the live view screen and the important functions you'll be using during flight.

1. **Live View Screen:**

    The live view is the primary screen where you can see the footage from the drone's camera in real-time. This screen provides several important data points and controls at a glance.

Camera Feed: At the center of the screen, you'll see a live feed from the drone's camera. This is your view of the world below as you fly.

Battery Status: A small battery icon on the top left will show the remaining battery life for both the drone and the controller.

GPS Signal Strength: On the top right, you'll see the number of satellites your drone is connected to. A higher number of satellites provides better stability and positioning for the drone.

Altitude and Distance: In the top-center area of the screen, you'll find your drone's current altitude and horizontal distance from the home point.

Heading/Orientation: The app displays the drone's heading and orientation, which helps you navigate during the flight.

2. **Key Functions and Features:**

Flight Mode Selector: On the left side of the screen, you'll find a flight mode button. Here, you can switch between different flight modes, such as Normal, Sport, and Tripod. Each mode has different speed and maneuverability settings.

Camera Settings: At the bottom right of the screen, you'll see a camera icon. Tap this icon to access various camera settings, such as resolution, frame rate, white balance, and exposure. This is where you can fine-tune your camera for optimal video and photo quality.

Telemetry Data: Telemetry provides real-time data about your flight, including speed, altitude, battery life, and GPS signal strength. This information is displayed in the top-left and bottom-right corners of the screen.

Flight Path and Map: On the lower-left corner, there's a map showing the position of your drone in relation to your home point, which helps you track your flight and return to base if necessary.

The **DJI Fly app** has a comprehensive menu that allows you to customize settings, adjust camera controls, and enable important safety features.

Here's an overview of what you'll find in the app's settings menu:

1. **Camera Settings:**

    In the camera settings section, you can change a variety of parameters related to your drone's camera, such as:

    Video Resolution: Choose the desired resolution (4K, 1080p) and frame rate (24fps, 30fps, etc.) for your video recording.

    Photo Settings: Adjust settings like white balance, ISO, shutter speed, and exposure compensation for the perfect shot.

    Grid Lines and Safety Lines: Enable grid lines on the live view to help with composition and horizon leveling. You can also enable safety lines to indicate your drone's distance from the home point.

2. **Safety Features:**

Obstacle Avoidance: The DJI Air 3S comes with advanced obstacle sensors, and you can enable or disable obstacle avoidance in the settings menu. This ensures that the drone automatically detects and avoids obstacles during flight, enhancing safety.

Return-to-Home (RTH): The RTH function is a critical safety feature that automatically returns your drone to its takeoff point when the battery is low or the signal is lost. You can configure the altitude and behavior of the RTH function in the settings.

Geofencing and No-Fly Zones: The app allows you to customize the geofencing settings, including no-fly zones, based on local regulations or personal preferences. This is especially important to avoid restricted airspace or areas near airports.

3. **Flight Settings:**

Flight Mode Customization: You can adjust flight modes (e.g., Normal, Sport, and Tripod) in the settings menu, allowing you to tailor your drone's flight characteristics to your needs.

Gimbal Settings: The DJI Air 3S features a gimbal for stable footage, and you can adjust the gimbal's tilt speed, smoothness, and other parameters through the settings.

Auto Takeoff/Landing: Enable or disable automatic takeoff and landing features, which simplify the flight process for beginners.

4. **Other Settings:**

   Firmware Updates: The DJI Fly app will notify you when a firmware update is available for your drone or controller. Always ensure that both devices are updated to the latest version for optimal performance.

   Camera Calibration: In the app, you can calibrate the camera and gimbal for more accurate shots. This can be helpful if you notice the camera isn't leveling correctly or is behaving erratically.

Wi-Fi and Connectivity Settings: Here, you can manage the Wi-Fi connection between your controller and mobile device, ensuring a stable link for the best user experience.

The DJI Fly app provides several options for personalizing your experience. You can adjust the display layout, toggle sound notifications, or switch between imperial and metric units for telemetry data. By exploring the app's menu, you can tailor it to suit your specific needs, making the flight experience more intuitive and enjoyable.

Chapter 3: Flight Modes and Controls

The **DJI Air 3S** offers three primary flight modes—**Normal**, **Cine**, and **Sport**—each designed to suit different flying needs and experiences. Whether you're a beginner still getting the feel for your drone or an experienced pilot looking for speed and precision, understanding how to switch between these modes can dramatically improve your control and safety during flights. Let's dive into each mode and see when and why you should use them.

In **Normal Mode**, the drone operates in a balanced, stable manner. This is the go-to mode for beginners and casual drone enthusiasts who are just getting started or want a smooth, easy flight experience. The control sensitivity is moderate, and the drone responds predictably to inputs, making it easier to fly steadily. When you're in Normal Mode, the drone's obstacle sensors are active, providing added safety, especially in environments where there may be obstacles. It's ideal for general use, where you just want to enjoy the experience of flying or capture basic aerial footage without worrying too much about quick maneuvers or sharp turns. This mode ensures that the drone won't suddenly take off too fast or make unpredictable movements, making it easier to get used to the controls and gain confidence in the air.

Cine Mode, on the other hand, is tailored for smoother, cinematic flying. When you switch to this mode, you'll notice that the controls are less responsive and more sluggish. While this may feel strange if you're used to more agile flying, this intentional sluggishness allows for more stable and controlled movements, which is perfect when capturing high-quality, stable video footage. The slow response time makes it easier to create fluid, cinematic shots without jerky or erratic motions, which is a key factor in professional drone videography. **Cine Mode** is perfect for capturing sweeping aerial footage, following a subject slowly, or getting those smooth fly-throughs and gentle rises or descents. If you're planning on creating videos where stability is paramount, **Cine Mode** is the way to go.

Lastly, there's **Sport Mode**, which is the fastest and most responsive of the three. This mode is for experienced pilots who want the freedom to push their drone to its limits. In Sport Mode, the drone becomes more agile, with sharper, quicker movements and higher speeds. If you're practicing advanced aerial tricks or racing your drone through tight spaces, this is the mode you'll want to use. However, because the drone responds so quickly and aggressively, you'll need to be fully comfortable with the controls. Sport Mode is not ideal for beginners, as it requires quick reflexes and good coordination to maintain control, especially in challenging or high-speed scenarios.

Switching between these modes is as simple as tapping the flight mode selector on the DJI Fly app. Knowing when to use each mode is key to optimizing both your control and safety. **Normal Mode** is great for starting out or flying in stable conditions, **Cine Mode** is the best for capturing professional-looking footage, and **Sport Mode** is perfect for those seeking high-speed thrills or advanced maneuvers. Having this knowledge in hand will help you master the air and get the most out of your DJI Air 3S.

Flying a drone like the **DJI Air 3S** requires a solid understanding of how to operate the remote controller, as it's your primary tool for controlling the drone's altitude, direction, and camera adjustments. The remote controller features an ergonomic design, with all the necessary buttons and joysticks placed intuitively for easy use. Let's break down how to master the remote and take full control of your drone.

The **left joystick** controls the altitude and yaw (rotation) of the drone. Pushing the joystick forward causes the drone to ascend, while pulling it backward brings the drone down. To rotate the drone, simply move the joystick left or right. The **right joystick**, on the other hand, controls the drone's pitch (tilt) and roll (side-to-side movement). Moving the right joystick up or down will tilt the camera to capture shots at different angles, while left or right movements of the joystick control the drone's direction of flight. These basic controls are the building blocks for any drone pilot, and understanding how to use them in combination is key to maintaining steady flight.

To **take off**, place the drone on a flat surface and power on the drone and controller. Slowly push the left joystick forward until the drone begins to lift off the ground. You'll feel the drone's motors humming as they take on the load of the airframe, but don't push the joystick too aggressively, as a gentle takeoff helps maintain stability and avoids any jerky movements. Once the drone is a few feet above the ground, hold the joystick steady to maintain a consistent altitude.

When it's time to **land**, simply lower the left joystick slowly. The drone will descend gradually. Be mindful of the drone's speed and ensure that the landing area is clear of obstacles. If you're landing in a confined space, you might need to use the right joystick to adjust the direction, ensuring that you're coming down smoothly and directly over your intended landing spot. The drone's obstacle sensors will help avoid any collisions as it nears the ground.

Mastering basic **directional movements** is key for beginner pilots. Flying in a straight line requires subtle adjustments to both the left and right joysticks. For smooth, controlled flight, practice pushing the joysticks gently in short bursts to get a feel for the sensitivity. The drone will naturally try to keep a stable position, but you can use the joysticks to move it in any direction you desire, whether it's forward, backward, or sideways.

Once you're comfortable with hovering, takeoff, and basic directional control, it's time to progress to more advanced maneuvers. Flying in circles is a great exercise to test your control over both the altitude and direction of the drone. Start by flying in a straight line, then gently adjust the right joystick in a circular motion to make the drone turn. Keep your altitude steady with the left joystick to maintain a consistent circle.

Another advanced maneuver is **tracking a subject**. In this case, the drone will follow a moving object or person, keeping them in the center of the frame while adjusting the drone's flight path as needed. This is often used for videography to follow a person, car, or animal from above. To track a subject, activate the **ActiveTrack** feature in the DJI Fly app, select your subject, and the drone will automatically adjust its position and speed to keep up with them.

For even more advanced skills, try performing a **figure-eight** maneuver. This involves flying the drone in a continuous loop—first in one direction, then reversing to form the second loop. This maneuver challenges your ability to control both altitude and direction, and it's an excellent way to test your coordination while also practicing smooth transitions between different flight paths.

These maneuvers—hovering, takeoff, landing, flying in circles, tracking, and performing figure eights—are the foundational skills that will help you gain confidence in your ability to pilot the

DJI Air 3S. With practice, you'll be able to take on more complex flights and shoot more dynamic, cinematic footage. Just remember, every flight is a learning experience, so take your time, keep practicing, and enjoy the incredible freedom that comes with flying a drone like the Air 3S.

Chapter 4: Camera Settings and Filming Techniques

The **DJI Air 3S** is not just a marvel of engineering when it comes to flight performance; it's also packed with a powerful camera system that is capable of capturing stunning aerial footage with remarkable clarity and detail. Let's dive deep into the camera features and how to adjust them to suit your needs.

At the heart of the **DJI Air 3S** is its 4K camera, which provides exceptional video quality at up to **60fps**. This resolution ensures that every shot is crisp, clear, and suitable for professional-grade video production. Whether you're shooting in broad daylight or at dusk, the Air 3S delivers vibrant and sharp images. With **HDR (High Dynamic Range)** capabilities, you'll experience rich color contrasts and deep shadows without losing detail in the brightest areas of the frame. The ability to shoot at 60 frames per second means you can capture smooth, fluid motion, making it ideal for action-packed scenes or fast-moving subjects.

One of the standout features of the **Air 3S** is its **3x telephoto lens**. This allows you to zoom in on distant objects without sacrificing image quality. Whether you're trying to capture a distant mountain range, a wildlife shot, or architectural details from a height, the telephoto lens provides a unique perspective that a traditional wide-angle lens cannot. It opens up new possibilities for capturing faraway subjects while maintaining the sharpness and detail you'd expect from a high-end drone camera.

When it comes to **adjusting camera settings**, the **DJI Fly app** gives you full control over everything from ISO sensitivity to shutter speed. In bright, sunny conditions, you may want to reduce the ISO to keep your images from becoming overexposed. In low-light settings, you can increase the ISO slightly, though you'll want to avoid pushing it too high to minimize noise. For video, setting the shutter speed to **double the frame rate** is a great rule of thumb for smooth, cinematic footage. For instance, if you're shooting at 24fps, you'd set your shutter speed to **1/50** to achieve natural-looking motion blur.

In addition to these manual settings, the **DJI Air 3S** offers automated modes that can adapt to your environment, helping you capture stunning shots without having to fiddle with settings in real-time. If you're shooting a sunset, for example, the **Auto-Exposure Bracketing (AEB)** function can automatically capture multiple exposures, allowing you to select the best one. Whether you're in a bright, sunny environment or shooting at dusk, the Air 3S offers enough flexibility to handle a variety of shooting situations with ease.

Capturing smooth, cinematic footage with the **DJI Air 3S** requires a combination of understanding your drone's camera features and applying best practices to your filming techniques. To achieve the most professional-looking video, **stabilization** is key. The **DJI Air 3S** uses advanced **3-axis mechanical gimbal stabilization**, which ensures that even in windy conditions or during fast movements, the footage remains steady and smooth. This is particularly important for filming dynamic, action-based scenes or tracking subjects.

When shooting video, one of the first things you need to think about is **framing**. A good shot begins with proper composition. The **rule of thirds** is a helpful guide—imagine a grid dividing your screen into nine equal parts with two horizontal and two vertical lines. Place your subject at one of the intersections for a more visually appealing shot. Additionally, consider the **leading lines** in your environment. Roads, rivers, or even mountain ridges can serve as natural guides to direct the viewer's attention and add depth to your shots.

Focus is another crucial aspect of filming. While drones like the Air 3S offer automatic focus, it's always a good idea to manually adjust the focus when filming close-up shots, especially if your subject is moving quickly. For **wide shots**, a broader depth of field will keep everything in focus, but for **close-ups**, you'll want to focus directly on your subject to avoid distracting background blur.

Lighting also plays a huge role in the quality of your footage. **Golden hour**, which occurs during the first hour after sunrise and the last hour before sunset, is ideal for shooting drone video. The soft, diffused light creates dramatic, rich scenes with fewer harsh shadows. If you're filming at midday when the sun is high, you may find that shadows can be too harsh, making it harder to capture smooth, natural footage. To counteract this, you might want to adjust the camera settings to **reduce the exposure** or use a **ND filter** (Neutral Density filter), which acts like sunglasses for your camera, allowing you to film in bright light while still maintaining the ideal shutter speed.

For **still images**, the same principles apply. Proper framing and composition are essential, but you also need to pay close attention to your drone's settings, particularly exposure. Overexposure can easily ruin a photo, and underexposure can lead to dark, unusable images. Again, experimenting with ISO settings, shutter speed, and aperture will help you achieve the perfect shot depending on the environment you're flying in.

The **DJI Air 3S** doesn't just give you control over manual settings; it also comes with a host of **intelligent flight modes** that take much of the guesswork out of capturing stunning shots. These modes are perfect for both beginners and seasoned pilots who want to achieve cinematic shots with minimal effort.

Mastershots is one of the most powerful features in the Air 3S's toolkit. This intelligent mode allows the drone to automatically fly a pre-programmed flight path while capturing a series of shots, all designed to give you professional-looking results. With just a tap of a button, the drone will execute complex maneuvers like **spirals**, **orbits**, and **ascending shots** while keeping your subject in focus. Whether you're shooting a landscape, an event, or a subject in motion, Mastershots makes it easy to create dynamic and captivating sequences.

Quickshots is another fantastic tool, designed for those who want a quick, easy way to capture cinematic footage with minimal input. With Quickshots, the drone automatically flies in specific patterns around the subject—**Dronie** (pulling away from the subject), **Circle** (orbiting around the subject), **Helix** (ascending while circling the subject), and more. Each mode creates visually stunning video that looks like it was taken by a professional drone operator, without requiring you to manually control every aspect of the flight.

For those who want to capture impressive time-lapse footage, **Hyperlapse** is the mode to use. This feature allows you to shoot a time-lapse with the drone in motion, which gives the effect of speeding up time while keeping everything stable and smooth. Hyperlapse can be used in various environments, whether you're filming a city skyline, a moving subject, or a natural landscape.

Lastly, **Panorama** is perfect for capturing wide, breathtaking landscapes. With a few taps, the drone will automatically take a series of shots and stitch them together into a wide-angle panoramic image. This is particularly useful for capturing sweeping vistas where a standard shot just doesn't do justice to the vastness of the landscape.

By combining these intelligent modes with the camera's high-quality features, the **DJI Air 3S** offers a range of tools that make it easier than ever to create breathtaking footage with just the push of a button. Whether you're looking for automatic cinematic sequences or you want to try your hand at professional-level shots, these modes take your creativity to new heights.

Chapter 5: Understanding GPS and Satellites

The **DJI Air 3S** relies heavily on **GPS technology** for precise navigation and safety during flight. GPS, or **Global Positioning System**, allows your drone to determine its exact location on Earth by connecting to satellites orbiting in space. This is essential for not only guiding the drone through its flight path but also for ensuring that it stays in the air safely and returns home if necessary.

When you first power up your drone, it begins searching for available satellites in the sky. The **GPS system** uses data from these satellites to triangulate the drone's position and maintain its stability in the air. This process is vital for ensuring that the drone stays within a defined flight area and doesn't drift off-course due to wind or other external factors. The more satellites your drone connects to, the more accurate its positioning will be, and the safer the flight becomes.

Before taking off, it's essential to **set up the home point**, which is the location where your drone will return if it loses connection to the controller or if you manually trigger the **return-to-home (RTH)** function. To do this, ensure that your drone is powered on and has connected to at least **12-13 satellites**. This is the minimum number required to establish a reliable GPS connection and ensure that the home point is set accurately.

Satellite connection is an integral part of the DJI Air 3S's functionality. Once you power on the drone, it should automatically search for satellites and establish a connection. The system typically shows you the number of satellites the drone is currently connected to via the **DJI Fly app**. Ideally, your drone should have at least **12-13 satellites** before you take off for safe flight operations. This is critical for enabling the **return-to-home** feature, which automatically directs the drone to return to its takeoff point in the event of signal loss or low battery.

At times, you may encounter issues with satellite connection, which can affect your drone's stability and performance. Some common problems include **poor GPS signal** in areas with tall buildings, dense forests, or locations with electromagnetic interference. In such cases, the drone might struggle to lock onto enough satellites, leaving you with fewer than the recommended 12-13 connections.

If you find that the GPS signal is weak or if the app indicates fewer than 12 satellites, there are a few things you can do to troubleshoot the issue. First, **move to an open area** away from obstacles like buildings, trees, or power lines, as these can obstruct the satellite signals. If you're flying indoors or under dense canopy cover, the signal will be significantly weaker, making it harder for the drone to get a solid GPS lock.

Another solution is to **wait for a few minutes** while the drone continues to search for a more stable connection. Sometimes, the system just needs a bit of time to acquire more satellites and adjust its position. If you're still facing issues, you may need to **reboot** the drone and controller and try again. Occasionally, a software update may also be necessary, so make sure your drone's firmware and the DJI Fly app are both up-to-date.

If you're in an area where satellite signals are limited, it's advisable to avoid flying until the connection improves. Flying with fewer than 12 satellites can cause the drone to behave erratically, and the **return-to-home function** may not work as expected. If the drone can't establish a proper GPS connection, it might not be able to accurately return to its home point, putting your flight at risk.

In cases where GPS is unavailable or unreliable, **the DJI Air 3S** will rely on **vision positioning** and other sensors to stabilize itself. However, GPS is always the most reliable method for accurate navigation and safety, so it's critical to ensure that you have a strong satellite connection before you take off.

In summary, having a solid GPS connection is vital for your drone's safety and operational success. Always check the number of satellites in the DJI Fly app before starting your flight, and take the time to ensure that you are in an area where the drone can connect to at least 12-13 satellites. By doing so, you'll maximize the safety and performance of your DJI Air 3S, ensuring that you get the best flying experience possible.

Chapter 6: Obstacle Avoidance and Safety Features

The **DJI Air 3S** is equipped with advanced **obstacle avoidance** technology, which is crucial for ensuring a safe and smooth flight experience. The drone utilizes an array of **sensors and cameras** to detect and avoid obstacles in its flight path. These sensors work in tandem with the onboard **vision positioning system**, creating a comprehensive layer of safety that helps the drone navigate around potential dangers like trees, buildings, or other objects in its path.

The drone is equipped with **forward, backward, and downward sensors**, which constantly scan the environment for obstacles. The **forward sensors** are primarily responsible for detecting objects in front of the drone, while the **backward**

sensors monitor the area behind it. The **downward sensors** help the drone maintain its position when flying at lower altitudes and prevent it from accidentally crashing into the ground.

The onboard **cameras** also play a crucial role in obstacle detection, providing real-time imagery to the sensors. Together, the sensors and cameras can detect obstacles from a range of distances, giving the drone ample time to react. This system works in **real-time**, constantly updating as the drone moves through its environment, making split-second decisions to help avoid collisions.

In the **DJI Fly app**, you can access and adjust the obstacle avoidance settings to suit your flying needs. The three primary modes for obstacle avoidance are **Bypass**, **Brake**, and **Off**.

1. **Bypass Mode**: This mode allows the drone to automatically avoid obstacles by flying around them. When the sensors detect an obstacle, the drone will **change direction** and find a way around it. This mode is ideal for flying in open spaces where there is ample room for the drone to maneuver.
2. **Brake Mode**: In this mode, if the drone detects an obstacle, it will **immediately stop** its forward motion. This is useful when flying in tighter spaces or when you need to ensure that the drone doesn't accidentally crash into something in front of it.
3. **Off Mode**: This mode disables the obstacle avoidance system, meaning the drone will not detect or avoid obstacles. It's generally

not recommended for beginners, as it places the responsibility of navigation entirely on the pilot. However, in some situations, experienced pilots may choose to turn it off when flying in environments where the system could misinterpret objects or where manual control is required.

In addition to obstacle avoidance, the **DJI Air 3S** is equipped with a range of **safety features** that work together to enhance flight security and give pilots peace of mind. Some of these features include **geofencing**, **no-fly zones**, and **automatic return-to-home** (RTH).

- **Geofencing**: This feature uses GPS data to create virtual boundaries that restrict the drone's flight in certain areas. For instance, the drone will be unable to fly in restricted or dangerous zones, such as airports, military areas, or sensitive government locations. This is an important feature that prevents the drone from accidentally entering no-fly zones and violating local regulations.

- **No-Fly Zones**: Similar to geofencing, no-fly zones are predefined areas where the drone is not permitted to fly. These zones are typically areas with high air traffic or other aviation concerns. The DJI Fly app provides real-time notifications about no-fly zones, ensuring that you are always aware of restricted areas.

- **Automatic Return-to-Home**: This feature ensures that your drone returns to its takeoff point if it loses signal, encounters low battery, or experiences other flight anomalies. The **RTH** function is a crucial safety feature, and it can be manually activated at any time through the app. When activated, the drone will autonomously fly back to its home point, avoiding obstacles and ensuring that it lands safely.

Additionally, the DJI Air 3S has built-in **altitude limits** to prevent the drone from flying too high, especially in areas with airspace restrictions. The system also provides **battery warnings** when the power is running low, giving you ample time to bring the drone back before it runs out of charge. In the event of a power loss, the **fail-safe mechanisms** ensure that the drone will either land safely or return home automatically.

While the DJI Air 3S is packed with safety features, it's still important to follow best practices when flying, especially if you're new to drone flying. Here are some practical tips to help you minimize risks and stay safe:

1. **Always Pre-Flight Check**: Before every flight, ensure that your drone, controller, and the **DJI Fly app** are functioning properly. Check the battery levels of both the drone and controller, and ensure that the GPS signal is strong (at least 12-13 satellites). Inspect the propellers and other components to make sure they are in good condition.

2. **Fly in Clear, Open Spaces**: As a beginner, it's best to fly in wide open areas with plenty of room to maneuver. Avoid flying in tight or cluttered spaces where obstacles might be harder to detect. Open fields or large parks are ideal for first-time flights.

3. **Weather Considerations**: Always check the weather before flying. Avoid flying in **strong winds**, heavy rain, or poor visibility conditions. These weather conditions can significantly affect the drone's stability and performance. It's also advisable to avoid flying during extreme temperatures, as they can impact battery life and drone performance.

4. **Respect the Environment**: When flying in natural environments, avoid disturbing wildlife, especially in areas that are known for nesting birds or other animals. Respect local regulations and guidelines for drone usage, and avoid flying too close to people or private properties.

5. **Always Keep the Drone in Line of Sight**: It's important to always have visual contact with your drone during flight, even if

you're using the camera to film. This ensures that you can react quickly in case of an emergency or if the drone encounters an obstacle that the sensors cannot detect.

By following these best practices and taking advantage of the safety features built into the DJI Air 3S, you can minimize risks and have a safe and enjoyable flying experience. Whether you're a seasoned pilot or a first-time flyer, being proactive about safety ensures that you get the most out of your drone without compromising on security.

Chapter 7: Advanced Flight Techniques

As you become more comfortable with your DJI Air 3S, the next step is to elevate your flying skills and master advanced maneuvers that can help you capture stunning, professional-quality footage. The drone's capabilities are vast, and with practice, you can push its limits to create truly dynamic and breathtaking shots.

Flying in tight spaces is a skill that requires precision and control. Whether you're weaving between trees, flying through narrow corridors, or navigating obstacles in a confined area, your DJI Air 3S can handle it with ease, provided you maintain steady control. The key to mastering these tight maneuvers is to stay calm, take it slow, and rely on the drone's obstacle avoidance system. Although the sensors will help you navigate around objects, it's important to gauge your speed and altitude carefully, ensuring that you don't overwhelm the system. When flying in tight spaces, using the **Cine Mode** can help you maintain a smooth, steady flight path, as it reduces the drone's speed and makes it easier to avoid sudden obstacles.

Capturing **high-speed action** is another exciting challenge. Whether you're following a fast-moving subject like a car, athlete, or animal, your drone's **Sport Mode** is ideal for this type of filming. In Sport Mode, the drone responds to commands more quickly and allows you to fly at higher speeds. It's important to practice your control techniques before attempting high-speed shots, as sudden jerks or rapid turns can cause instability or blur the footage. To ensure smooth footage during fast flights, anticipate the movements of your subject and plan your shots accordingly. High-speed filming often requires **quick reflexes**, so familiarize yourself with the responsive controls and the drone's behavior in Sport Mode before capturing action-packed scenes.

Aerial photography presents its own set of challenges and opportunities. With the DJI Air 3S, you have a powerful tool at your disposal for capturing stunning landscapes, architecture, and outdoor scenes. The drone's **4K camera** and **3x telephoto lens** give you exceptional clarity and flexibility in framing your shots. When setting up for aerial photos, it's important to consider the **composition**. Use the rule of thirds to create a balanced and visually appealing image. Try to frame the shot by placing key elements along the gridlines or intersections for maximum impact.

One of the most powerful aspects of the DJI Air 3S is its **intelligent flight modes**, which can help you create dynamic, cinematic shots without needing to manually control every aspect of the flight. **Mastershots** is one such mode that allows the drone to automatically perform a series of pre-programmed maneuvers while capturing footage. The drone will intelligently frame the subject and adjust its flight path to create a sequence of smooth, cinematic shots that can be edited together for a stunning video. Whether it's a dramatic reveal shot or a series of sweeping aerial views, Mastershots can take your filming to the next level.

Similarly, **Quickshots** is a great option for capturing dynamic, high-energy shots in a short amount of time. It offers a variety of preset flight paths and camera movements, such as the **Dronie**, **Circle**, **Helix**, and **Rocket**, each designed to add creative flair to your video. These modes are perfect for capturing subjects on the move or for adding cinematic effects to your shots with minimal effort.

The **Hyperlapse** mode is a fantastic tool for creating time-lapse sequences, especially when you want to show a passage of time in a unique, high-speed format. You can select from a variety of preset paths, or even customize your own, to create stunning time-lapse footage over long distances. For example, you can capture the movement of clouds, the changing light of a sunset, or the shifting landscape in a breathtaking way. Just make sure that your battery life is sufficient to complete the entire time-lapse sequence, as these shots often require longer flight times.

Lastly, the **Panorama** mode is an excellent choice for capturing expansive landscapes and wide-angle scenes. This mode automatically stitches together a series of photos to create a high-resolution panoramic image. The drone will take a series of shots from different angles and heights, which it will then merge into one cohesive image. This is perfect for wide vistas, mountain ranges, and sweeping cityscapes.

Filming in **challenging environments**—such as low light, windy conditions, or other difficult settings—requires both technical understanding and creative adaptation. The DJI Air 3S is equipped to handle these challenges, but knowing how to adjust your settings and fly in tough conditions is crucial for achieving the best results.

In **low-light situations**, such as during sunrise or sunset, you'll need to adjust your camera settings to ensure the image remains clear and vibrant. The drone's camera has built-in low-light capabilities, but it's important to manually adjust the **ISO** and **shutter speed** for optimal exposure. A higher ISO will allow more light to be captured, but it may introduce noise or grain into the footage. A lower ISO will give you cleaner images but may darken the scene. Finding the right balance is key. Similarly, adjusting the **shutter speed** can help maintain the fluidity of motion while avoiding blurry footage. A good rule of thumb is to set the shutter speed to **double the frame rate** for smooth cinematic shots.

Flying in **windy conditions** can be challenging, as strong gusts can make the drone harder to control. However, the DJI Air 3S is equipped with an advanced stabilization system that helps keep the drone steady in the air, even when winds pick up. To get the best footage in windy conditions, try to fly at lower altitudes, where the wind is less turbulent. Always monitor the drone's battery levels closely when flying in challenging weather, as the wind can cause the drone to work harder and drain the battery faster.

There are also situations where you'll need to **adapt to your drone's limitations**. While the DJI Air 3S is a powerful drone, it has its constraints, particularly when it comes to battery life, range, and environmental factors. To overcome these limitations and capture professional-quality footage, it's important to plan your flights carefully. Monitor your battery levels and ensure you have enough power to complete your shot or return safely. Avoid flying too far from your home point, as the drone's range is limited, and you don't want to risk losing the connection.

In more extreme environments, such as mountainous terrain or over water, always be aware of potential hazards. Water and drones don't mix well, so avoid flying over large bodies of water unless you're confident in the drone's waterproofing features (though the DJI Air 3S is not fully waterproof, so it's best to play it safe). Additionally, keep an eye on any obstacles like trees, power lines, or buildings, which can interfere with the drone's signal and flight path.

By understanding your drone's capabilities and limitations and mastering advanced flight techniques, you can capture breathtaking footage, even in the most challenging environments. Whether you're flying through narrow gaps, capturing fast action, or filming in difficult weather, the DJI Air 3S gives you the flexibility to create professional-quality aerial footage with ease.

Chapter 8: Drone Maintenance and Troubleshooting

To ensure that your DJI Air 3S continues to perform at its best, routine maintenance is essential. Taking proper care of your drone will not only prolong its lifespan but also ensure that it remains safe and reliable for every flight.

Routine maintenance tasks should start with regular **propeller checks**. Propellers are the most exposed part of your drone, and they can easily become damaged if the drone crashes or comes into contact with objects. Inspect the propellers before every flight, looking for signs of wear, cracks, or chips. If any damage is noticeable, replace the propeller immediately to avoid potential flight issues. Additionally, always ensure that the propellers are securely attached before takeoff. Loose or improperly installed propellers can cause instability, making it difficult to control your drone.

Battery care is another crucial aspect of maintaining your drone's performance. The battery is the heart of the DJI Air 3S, and taking care of it will help extend its life and ensure that you get the most out of every flight. Always make sure that you store the battery in a cool, dry place, away from extreme temperatures, as excessive heat or cold can negatively affect its performance. Avoid letting the battery fully discharge, as this can reduce its lifespan. Ideally, keep the battery charge between 20% and 80% for long-term storage. When charging, never leave the battery unattended, and always use the provided charger to avoid damaging the battery or causing a safety issue.

To ensure your drone is running the latest software and taking advantage of any improvements or bug fixes, **firmware updates** are vital. DJI frequently releases firmware updates that improve flight performance, enhance camera features, and fix any known issues. These updates can be downloaded directly from the DJI Fly app, and it's a good idea to check for updates before every flight, especially if the drone hasn't been flown in a while. Follow the instructions carefully when updating the firmware, and always ensure that the drone is fully charged before starting the process.

When it comes to **storing your drone**, always make sure to protect it from physical damage. Store the drone in a carrying case or a protective pouch to keep it safe from scratches, dust, or impacts when not in use. Make sure that the drone is powered off and that the propellers are removed if storing it for an extended period. If you're traveling with your drone, use a dedicated backpack or case designed for drones to protect it during transport. Additionally, avoid leaving your drone in places where it could get exposed to moisture, like in a car on a hot day or in a damp environment.

In addition to regular maintenance, it's important to know how to troubleshoot common issues that may arise during your drone flights. **Connectivity issues** are one of the most common problems drone pilots face, and they can usually be traced to either the remote controller or the drone itself. If you're experiencing a weak or intermittent signal, ensure that the controller is fully charged and that there are no obstructions between you and the drone. Check that the drone is connected to the correct frequency and that no interference is coming from nearby electronic devices. If the issue persists, try restarting both the drone and the controller, and reconnect them using the DJI Fly app.

Camera malfunctions can also be frustrating, especially if you rely on your drone for aerial photography or videography. Common camera issues include blurry or out-of-focus footage, exposure problems, or the camera freezing. If the camera isn't focusing properly, check that the lens is clean and free from dirt or debris. In some cases, recalibrating the camera's gimbal might be necessary to restore stability and ensure smooth footage. For exposure problems, check your settings within the DJI Fly app and ensure that the camera is set to the appropriate ISO and shutter speed for the lighting conditions. If the camera freezes or fails to operate, a simple restart of the drone and app might solve the issue.

GPS signal loss can happen, particularly in areas with poor satellite coverage, such as dense forests or urban environments with tall buildings. If you experience GPS signal loss during flight, the first thing to do is ensure that the drone has a clear line of sight to the sky. Try moving to an area with fewer obstacles and wait for the drone to re-establish a connection to the satellites. In some cases, simply waiting for a few minutes or repositioning the drone may be enough to restore the GPS signal. However, if the problem persists, it may be a sign of a more significant issue, such as a malfunctioning GPS module or interference from nearby sources.

In the event that you encounter any problems that you cannot resolve on your own, **contacting**

DJI support is the next step. DJI offers excellent customer service and can help troubleshoot issues with your drone. Be ready to provide detailed information about the problem you're facing, including any error messages or unusual behaviors you've observed. If the issue is more complex and cannot be fixed remotely, DJI may direct you to a local **repair center**. It's important to always use official repair services to ensure that your drone is properly fixed and that the warranty remains intact.

By staying on top of routine maintenance and knowing how to troubleshoot common issues, you can keep your DJI Air 3S in optimal working condition and ensure that your flights are smooth, safe, and enjoyable. Regular care and attention will go a long way in prolonging the life of your drone, giving you many more hours of aerial adventures.

Chapter 9: Flying Legally and Responsibly

Understanding drone laws and regulations is crucial for every pilot, whether you're just starting out or have years of experience. As drone technology continues to advance, governments around the world are adapting their laws to ensure the safe and responsible use of drones in shared airspace. Ignorance of these regulations can lead to hefty fines, damage to property, or even the loss of your drone. Therefore, it's essential to familiarize yourself with the rules governing drone flight in your area and any region you plan to fly in.

A guide to drone laws varies by country and region, but there are common rules that most jurisdictions follow. In general, you need to be aware of **no-fly zones**, such as airports, military bases, and government buildings. These zones are designated to protect both the safety of aircraft and the privacy of individuals. Many countries also have specific regulations regarding **altitude limits**—for example, in the United States, the Federal Aviation Administration (FAA) restricts recreational drone flights to an altitude of 400 feet above ground level. Similarly, drones must typically remain within the operator's line of sight during flight, unless a special waiver or license is obtained.

Another key area of regulation concerns **distance limits**. In many countries, you are required to keep your drone within a certain range of the takeoff point, which may vary depending on local laws. For instance, in some areas, drones must stay within a 500-meter radius, while other regions may have stricter requirements. Be sure to check the specific regulations in your country or region before flying.

Privacy concerns are a significant issue in many places, and you must be mindful of the impact your drone could have on the privacy of others. Avoid flying over private property without permission, and be cautious when flying near residential areas. Recording or photographing people without their consent can lead to legal consequences. Always ensure that you are respectful of the people around you and mindful of where your drone's camera is pointed.

How to be a responsible drone pilot goes beyond simply following the law; it's also about demonstrating ethical behavior in your flying practices. For one, respecting privacy is of utmost importance. Flying your drone in a way that invades someone's personal space, or recording images or videos without permission, can lead to serious ethical concerns and legal action. Always ensure that you are flying in an area where you have permission to operate and avoid areas where people may reasonably expect privacy.

Additionally, you should be cautious when flying your drone near **people and animals**. Drones can be loud and intimidating, and they can cause anxiety or even harm to people or pets. Keep a safe distance from crowds, children, and animals, and be especially cautious in areas where there is wildlife. Many regions also have laws specifically prohibiting flying drones near certain types of wildlife, as it can disrupt their natural behavior and pose a danger to them.

Furthermore, you should always avoid **no-fly zones** such as airports, military areas, and other restricted airspace. Most countries provide maps of these zones, and drone operators are required to avoid them at all costs. For example, in the United States, the FAA mandates that drones stay at least five miles away from an airport without prior approval, while in Europe, airspace regulations similarly restrict drone flights in close proximity to airports and other sensitive locations.

Another important step in becoming a responsible drone pilot is **registering your drone**. In many countries, you are required to register your drone with the relevant aviation authorities. This helps ensure that drones are tracked and that pilots are aware of the rules and regulations that apply to them. For example, in the United States, the FAA requires drone registration for all drones weighing over 0.55

pounds (250 grams), while other countries have similar rules based on weight and purpose of use.

In some regions, you may also need to **obtain certifications** to operate your drone legally, especially if you plan to use it for commercial purposes. For example, in the United States, anyone flying drones for commercial purposes must obtain a remote pilot certificate from the FAA. Similarly, in Europe, the European Union Aviation Safety Agency (EASA) requires drone operators to pass an online training course for commercial flying. Additionally, many countries require drone pilots to pass safety tests or obtain specific licenses to fly drones that exceed certain weight limits or are used in more complex flying conditions.

By following these drone laws and being a responsible pilot, you can ensure that you are flying within legal boundaries and protecting the safety and privacy of those around you. Not only does this keep you safe from legal trouble, but it also helps the drone community as a whole maintain a good reputation. As drone use becomes more widespread, being aware of and adhering to local regulations will only become more important, ensuring a future where drones are flown safely and responsibly.

Chapter 10: Troubleshooting and Fixing Errors

Drone flying can be a rewarding and exhilarating experience, but like any technology, drones are susceptible to occasional errors and malfunctions. While some of these issues are simple to fix, others may require more careful troubleshooting. Being able to quickly identify and resolve common drone errors is an essential skill for all pilots. Here are some of the most common drone problems you may encounter, along with solutions and preventive measures to ensure a smooth flight every time.

Identifying and Resolving Common Drone Errors

1. **GPS Signal Loss**

    One of the most common issues drone pilots face is a loss of GPS signal, which can lead to erratic behavior or even a drone drift. This issue is often caused by poor satellite connection or interference from tall buildings, trees, or other obstacles.
    - **Solution**: First, ensure you're flying in an open area with minimal interference. If you encounter signal loss, try repositioning your drone to get a clearer line of sight to the sky. In some cases, moving to a higher altitude can help. If the problem persists, make sure your drone's

firmware is up to date, as updates often improve GPS performance.

- **Prevention**: Before each flight, check the satellite status on the DJI Fly app to confirm that your drone has connected to enough satellites (at least 12–13 for optimal performance). It's also recommended to wait a few moments before taking off to ensure the GPS lock is strong.

2. **Poor Video Quality**

Poor video quality is another frustrating issue, often linked to a weak signal or incorrect camera settings. This can result in grainy footage, pixelation, or laggy video feed.

- **Solution**: Check the connection between your drone and the remote controller. A weak signal can affect the quality of the live feed. Also, inspect your camera settings. Ensure that your resolution and frame rate are appropriately set for the conditions you're filming in. If you're flying in areas with high interference, consider using a different transmission channel or adjusting the video quality in the app.

- **Prevention**: Before flying, ensure that your drone's firmware is updated, as DJI often releases patches to improve camera performance. Also, check that the camera lens is clean and free of obstructions. If you're flying in an area with a lot of interference, try to minimize the use of other wireless devices or fly in areas known for stable transmission.

3. **Unresponsive Controls**

   At times, you may experience unresponsive controls, where your drone doesn't respond to your inputs. This could be caused by several factors, including connection issues, interference, or even problems with the controller itself.

   - **Solution**: Start by checking the connection between your drone and the remote controller. Ensure that both devices are fully charged and that you have paired them correctly. If the controls are still unresponsive, try restarting both the drone and the remote controller. A simple reset often resolves the issue. In rare cases, you may need to recalibrate the drone's sensors through the DJI Fly app.

- **Prevention**: Before each flight, always check the connection between the drone and the remote controller, and ensure both devices are properly paired. Make sure there are no obstructions or interference between you and the drone, especially when flying in urban areas or near large metal structures.

4. **Battery Issues**

Drone batteries are one of the most critical components, and any issues with battery life or charging can significantly impact your flight experience. Common problems include rapid battery depletion, charging errors, or the battery not holding a charge.

- **Solution**: If you experience rapid battery drain, check the battery's health in the DJI Fly app. If the battery is showing signs of deterioration, it may need to be replaced. If your battery is not charging properly, ensure that the charger and battery contacts are clean and free of debris. In some cases, a faulty charging cable or port can cause issues, so try using a different charging setup.
- **Prevention**: Always store your drone battery in a cool, dry place, and avoid overcharging it. When not in use, ensure that the battery is at around 50% charge for optimal

longevity. Be mindful of battery life during your flights, and try to land the drone before the battery hits critically low levels.

5. **Motor and Propeller Problems**

   If your drone starts making unusual noises or has trouble lifting off, the motors or propellers may be at fault. Damaged or misaligned propellers can prevent your drone from flying correctly, while motor issues can cause instability.

   - **Solution**: Inspect the propellers for any visible damage. If a propeller is cracked or bent, replace it immediately. Ensure that all propellers are securely attached and rotating freely. If the motors seem to be malfunctioning, check for dirt or debris that may be causing friction. If the problem persists, consult the drone's manual for guidance on motor calibration.

- **Prevention**: Before each flight, inspect the propellers and motors to ensure they are in good condition. Replace damaged propellers immediately and clean the motors as needed. Flying in a debris-free area can also help prevent damage to the propellers.

Preventive Measures

To minimize the risk of encountering issues during your flight, it's important to establish a routine for pre-flight checks. Here are some steps you can take before each flight to ensure your drone is in top condition:

1. **Check Battery Levels**

   Always ensure that both your drone and remote controller have sufficient charge before taking off. Avoid flying with a battery that's less than 20% charged, as this can reduce flight time and increase the risk of an unexpected battery drain.

2. **Inspect the Drone and Propellers**

   Inspect the drone for any visible signs of damage, such as cracks in the body, loose screws, or worn-out propellers. Check the propellers for any chips, cracks, or bends and replace them if necessary.

3. **Update Firmware**

   Make sure your drone's firmware is up to date before each flight. Firmware updates can improve drone performance, enhance safety features, and fix any known bugs or glitches.

4. **Calibrate the Compass and IMU**

   If you notice unusual behavior, such as drifting or unstable flight, recalibrate the compass and IMU (Inertial Measurement Unit) through the DJI Fly app. This will help ensure your drone is correctly aligned and ready for flight.

5. **Pre-Flight GPS Check**

   Ensure that your drone has a solid GPS connection before taking off. This will help ensure that your drone stays on course and that features like return-to-home work properly.

How to Reset the Drone and Remote Controller

If you encounter persistent issues that can't be resolved with basic troubleshooting, resetting the drone and remote controller can often solve the problem. Here's how you can do it:

- **Drone Reset**: To reset the DJI Air 3S, turn off the drone and remote controller. Press and hold the power button on the drone for about 10 seconds until the lights flash, signaling a reset. Once reset, power the drone back on and reconnect to the controller.
- **Remote Controller Reset**: To reset the remote controller, power it off, then press and hold the power button for 10 seconds. The remote controller will enter pairing mode, and you can reconnect it to your drone via the DJI Fly app.

By following these troubleshooting steps and preventive measures, you can significantly reduce the likelihood of technical issues during your drone flights, ensuring a safer, more enjoyable flying experience each time you take to the skies.

Chapter 11: The Future of Drone Technology

As the drone industry continues to evolve at a rapid pace, it's fascinating to think about what the future holds for both enthusiasts and professionals alike. DJI, a pioneer in the drone market, has been at the forefront of this revolution, consistently pushing the boundaries of what drones can do. The DJI Air 3S, as impressive as it is, represents only a part of the bigger picture — a snapshot of where the technology stands today and a glimpse of where it's headed in the coming years. The next few years promise to bring exciting advancements, with breakthroughs in camera technology, artificial intelligence (AI), and drone design likely to change how we interact with the sky.

What's Next for DJI Drones and the Drone Industry

Looking ahead, we can expect drone technology to continue its rapid advancement. One of the most exciting areas of development is camera technology. Drones, especially those used for aerial photography and videography, have become an essential tool for capturing stunning visuals. As consumer demands for higher-quality footage continue to rise, DJI and other manufacturers are developing cameras with even more impressive capabilities. The future of drone cameras is likely to feature higher resolutions, more powerful zoom lenses, and better low-light performance. Imagine being able to capture 8K video or 360-degree shots with the same ease as current 4K technology. The addition of HDR (High Dynamic Range) to cameras will improve color accuracy, allowing for even more cinematic results.

AI-powered flight is another area where drones are expected to see significant improvements. Right now, AI is already helping drones with intelligent flight modes like obstacle avoidance, subject tracking, and automatic return-to-home. In the future, we can expect these features to become even more sophisticated. For example, AI could allow drones to fly autonomously through complex environments, adjusting in real-time to changing obstacles and weather conditions. It might even become possible for drones to make decisions about the best camera angles or flight paths without any input from the pilot, creating a truly hands-off experience.

In terms of design, drones are becoming smaller, lighter, and more efficient. Battery life, one of the most critical aspects of drone technology, continues to improve, with new power systems offering longer flight times and quicker recharges. These advancements will make drones more accessible for a wider range of users and allow for more complex missions. With continued progress in battery and power management technology, we may eventually see drones capable of flying for hours at a time, rather than just the 30-40 minutes typical of today's models.

How the DJI Air 3S Fits into the Bigger Picture

The DJI Air 3S is a clear example of the incredible strides the drone industry has made in recent years. When compared to its predecessors, it offers a host of improvements that make it a versatile option for both beginners and seasoned pilots. From its superior camera capabilities, including the ability to shoot 4K footage at 60fps, to its enhanced flight modes and obstacle avoidance features, the DJI Air 3S represents a peak in consumer drone technology. But it's important to recognize that this model is part of a larger evolution in the drone industry.

The Air 3S is a direct result of the lessons learned from earlier DJI models, such as the Mavic Air and Mavic Pro. Each new model in the Air series has offered significant upgrades, and the 3S is no exception. DJI's commitment to improving flight stability, camera performance, and user experience shines through in this model. The introduction of new intelligent flight modes and an extended battery life further elevates the Air 3S to a level that's perfect for both casual pilots and professionals.

But this is only the beginning. The evolution of drone technology doesn't stop with the Air 3S. Future iterations of the Air series or entirely new drone models from DJI may include even more advanced features. We can expect drones to become more intuitive, with more AI-driven capabilities. The camera systems will likely evolve to include even more powerful sensors, providing better image quality across various conditions. We may also see more integration between drones and other technologies, such as 5G connectivity, to allow for faster, real-time data transmission, and enhanced remote control capabilities.

Predictions for Future Updates and Features in Upcoming DJI Models

As for the next steps in the evolution of DJI drones, the future looks incredibly promising.

One of the most anticipated advancements is the integration of 5G technology. With 5G, drones will be able to transmit high-definition video feeds in real time, even over long distances. This would allow for more dynamic, live-streamed drone footage, enhancing the capabilities of drones for events, news reporting, and other commercial uses.

Additionally, the use of machine learning and computer vision will likely become more prevalent. DJI drones could potentially use AI not just to avoid obstacles, but to understand and interpret the environment around them, adjusting flight patterns or camera settings automatically based on what it "sees." This could lead to drones capable of performing highly complex tasks, such as mapping out a site in 3D, tracking multiple objects simultaneously, or even identifying points of interest based on images and video.

Battery technology is also expected to take a leap forward. With the push toward greener technologies, we may see drones powered by more sustainable energy sources, offering longer flight times and faster recharge cycles. In addition, drones may feature smarter power management systems, using less energy when possible to extend flight time and reduce the environmental footprint.

Finally, it's likely that future DJI drones will become more customizable, allowing users to tailor their drone to their specific needs. Whether it's adding specialized sensors for certain types of filming or changing out the camera for a different model, the ability to personalize a drone will be an appealing option for many users, especially professionals who require specific capabilities for their work.

In conclusion, the DJI Air 3S fits perfectly into the broader context of drone development. It stands as a highly capable and feature-rich model, but it is also a glimpse into a future filled with even more advanced, intelligent, and powerful drones. As technology continues to evolve, DJI is likely to lead the charge in pushing the boundaries of what drones can do, giving us even more exciting possibilities for the future of aerial photography, filmmaking, and beyond. Whether you are a hobbyist or a professional, the sky's the limit, and with drones like the DJI Air 3S, we're only just beginning to see what's possible.

Conclusion

As you wrap up your journey with the DJI Air 3S, it's important to reflect on how far you've come and recognize the immense potential this drone holds for future adventures. From mastering basic flight maneuvers to capturing stunning footage and exploring advanced intelligent flight modes, you've laid the groundwork for a rewarding experience in the world of drone piloting. Maintaining and enhancing your skills is a continuous process, but with a bit of effort and curiosity, the possibilities are endless.

To keep your **DJI Air 3S** in top condition, always prioritize routine maintenance. Regularly inspect the propellers for wear, keep the camera lens clean, and charge your batteries properly to ensure optimal performance. Firmware updates are essential for accessing the latest features and improvements, so make it a habit to check for updates before each flight. Proper storage of your drone, especially during travel, is equally crucial to prevent damage. These small but vital steps will help ensure that your drone remains reliable and ready to perform when you need it.

Enhancing your flying skills comes down to consistent practice and a willingness to experiment. Try different flight modes, challenge yourself with complex maneuvers, and explore varied environments to push your abilities further. Focus on refining your control, whether it's perfecting smooth cinematic shots in **Cine Mode** or capturing fast action sequences in **Sport Mode**. The more comfortable you become with the controls and the drone's behavior, the more confident you'll feel when tackling ambitious projects or flying in challenging conditions.

Drone flying is an art as much as it is a skill. Every flight offers an opportunity to learn something new, whether it's about the drone's capabilities, environmental conditions, or your own creative style. Embrace these moments of discovery and don't shy away from challenges. Mistakes are part of the process, and every pilot, no matter how experienced, learns from them. Keep experimenting with new angles, lighting conditions, and camera settings to capture footage that tells a story or offers a fresh perspective. The beauty of drone piloting lies in its ability to let you see the world from a completely different vantage point—so keep exploring and pushing your creative boundaries.

To truly grow as a drone pilot, immerse yourself in the vibrant and supportive **drone community**. Joining forums, social media groups, or local drone clubs can connect you with like-minded individuals who share your passion. These communities are an invaluable resource for tips, tricks, and inspiration, and they can help you stay informed about the latest trends and updates in the drone world. Sharing your footage with others not only allows you to showcase your skills but also opens up opportunities for constructive feedback that can help you improve.

Consider attending **workshops and events** that focus on drone piloting and aerial photography. These sessions are often led by professionals who can provide hands-on guidance and insights into advanced techniques. Workshops also offer a great chance to network with other pilots and build relationships that may lead to collaborative projects or even professional opportunities.

If you're looking to take your skills to the next level, you might also explore **specialized certifications** or courses in drone piloting. Many organizations offer training programs that cover advanced topics like commercial flying, drone cinematography, and flight planning. Gaining certifications not only boosts your confidence but can also open doors if you're considering turning your passion into a career.

Finally, don't hesitate to share your work with the world. Platforms like YouTube, Instagram, and Vimeo are perfect for showcasing your footage to a global audience. Sharing your creations not only inspires others but also gives you a platform to document your progress as a drone pilot. You never know who might see your work—your next big opportunity could come from someone admiring your aerial shots.

In wrapping up your DJI Air 3S journey, remember that this is just the beginning. The skills and experiences you've gained with this drone are a foundation for a lifelong adventure in the skies. Keep practicing, keep experimenting, and most importantly, keep capturing the world from above. With dedication and creativity, the DJI Air 3S will continue to be your trusted companion in exploring new heights, discovering unseen perspectives, and sharing awe-inspiring stories. The sky is truly the limit—now it's time to make it yours.

www.ingramcontent.com/pod-product-compliance
Lightning Source LLC
Chambersburg PA
CBHW071652240526
45469CB00021B/2206